Air Plant Care and Design

Tips and Creative Ideas for the World's Easiest Plants

Ryan and Meriel Lesseig

Skyhorse Publishing

Skyhorse® and Skyhorse Publishing® are registered trademarks of Skyhorse Publishing, Inc.®, a Delaware corporation.

Visit our website at www.skyhorsepublishing.com.

10 9 8 7 6 5 4 3 2 1

Library of Congress Cataloging-in-Publication Data is available on file.

Cover design by Laura Klynstra
Cover photos by Ryan and Meriel Lesseig

Print ISBN: 978-1-68099-154-3
Ebook ISBN: 978-1-68099-156-7

Printed in China

Contents

This book, much like our plant business, has truly been a labor of love, and could not have been realized without the support of our family and friends. We dedicate this book to our parents, whose endless encouragement and love inspires us, lifts us up, and holds us together.

Preface: From Budding Curiosity to Full-Blown Air Plant Obsession

"How did you get into air plants?"

My husband and I hear this question all the time. Why in the world would two young-ish professionals, one with an established career in marketing (me) and one with a burgeoning career in real estate (him), start a company selling *Tillandsia*?

Let's start with a little personal history. Ryan was born and raised in Florida. He's always had a passion for horticulture and an aptitude for keeping plants alive and kicking. Even from a young age, as his mother would reveal to me with great affection, he enjoyed his "dirt time." He carried this love for flora and fauna into adulthood, and I first was able to truly appreciate this quality as he transformed the lackluster, tangled mess of weeds that surrounded our first house together into a beautiful, thoughtfully-tended tropical landscape.

I am so appreciative of his skill because, well, to say I have been cursed with a black thumb would be an understatement. I'm not quite sure why . . . the cards were stacked in my favor. I was raised in rural Vermont by parents who loved to garden, and were good at it. I grew up in a home filled with thriving houseplants, many of which my dad still tends to and several of which are older than me. And yet, even with great examples of agricultural prowess set by my parents, and although I have always loved and appreciated the beauty of plants, keeping them alive has never been my forte.

Fast forward to one fateful weekend trip to St Augustine, Florida. We're there to scope out wedding venues and enjoy a little getaway. We're strolling down St. George Street and spot a display of these curious little plants outside one of those seashell trinket stores that I normally avoid at all costs. But those plants—they are so cool! No soil needed? Hard for even *me* to kill them? We get a few home and we are in love. We want more. And we realize as we start digging around the internet, that there really aren't many online retailers selling them. We should change this, we say.

Fast forward a bit more and now, what started as a small side business to share our love of air plants with others has grown into a fulltime gig that's still growing. As our air plant business has grown, we've become more and more enthralled by these amazing plants—the variance in *Tillandsia* shape, their blooms, their resilience, and the endless ways to use them in design. One of the things that we love most about our business is the creativity of others that it has

exposed us to. We have had the privilege of working with incredibly passionate *Tillandsia* growers, insanely talented floral designers, and creative artisans who push the boundaries of air plant art and design. It is truly inspiring, and something we do not take for granted.

It is because of this—our love for each other, for air plants, and for the community that loves them too—that we have set out to make this book a collaborative journey. We hope that you will find it to be a helpful resource for *Tillandsia* education, but also that it will inspire you to find your own air plant creativity. We hope that you'll share our love for these amazing plants, and that you'll share this love with someone special to you who may be new to the wonderful world of *Tillandsia*.

We think that love and air plants are two things worth sharing.

A close-up of the bloom on the *brachycaulos x streptophylla* hybrid shows its lush red and multiple purple flowers.

Ryan & Meriel Lesseig are co-owners of Air Plant Design Studio, a premium online retailer of *Tillandsia*. They live in Tampa, FL with their two dogs Haley and Molly. Their mutual love of nature and design, and Meriel's strong background in marketing, lead them to develop their business, which they are continuing to grow both locally and nationally. They enjoy traveling, especially to the Florida Keys where they are able to combine their love of tropical plants with their love of the water.

All About Air Plants: *Tillandsia* Species and Care

Commonly know as air plants, *Tillandsia* is a genus within the Bromeliad family that consists of over 700 different varieties. While some *Tillandsia* varieties can be grown in soil, most prefer to be grown as epiphytes. Epiphytes are plants that grow on another host or object such as a rock or tree, and use the host only for support (not to derive nutrients from). Epiphytic *Tillandsia* thrive by collecting moisture and nutrients from trichomes that are located on their leaves. These unique plants come in many different shapes and sizes, with some

Tillandsia come in all shapes, sizes, and colors, and will feature distinctly different bloom types and cycles. Pictured here from left to right: a clump of blooming *Tillandsia "Houston"* produces bright pink flowers; a *bulbosa* variety blooms a vivid red inflorescence; a *Tillandsia Eric Knobloch* is a hybrid air plant with bright green leaves.

having very symmetrical leaf systems while others display wavy leaves that look almost alien in form. The structure of the plants can also widely vary, with some types having a more bulbous base and others more linear.

This *Tillandsia xerographica* cascades down an eclectically-styled book shelf. This mesic air plant enjoys bright sunlight and will have looser leaves the more hydrated it is, though it is one of the more drought-tolerant air plants

While commonly known and referred to as air plants, this term can be a bit misleading as not all air plants are *Tillandsia*, and many but not all *Tillandsia* species are epiphytic, the distinction that allows them to be referred to as "air plants."

Tillandsia species can be found in a wide range of environments throughout Mexico, Central America, South America, and even some parts of the Caribbean and southern United States. They are native to climates ranging from humid jungles to dry deserts, with some varieties even growing at higher elevations within the cloud forest. Most varieties will grow well in temperatures between 50 degrees and even into 90 degrees if given good airflow and frequent watering.

The majority of these species can be classified as either mesic or xeric. Mesic plants are generally greener, and prefer higher levels of humidity. Xeric plants are normally more silver-grey in color, can take higher levels of light, and thrive in drier conditions. Some highly drought-tolerant xeric varieties can even handle direct sunlight. Therefore, the type of *Tillandsia* should be considered seriously when selecting and caring for these plants.

Popular in the 1970s, *Tillandsia* have recently reemerged as an on-trend plant, due in large part to their relative ease of care. They are popular with tropical plant enthusiasts for their unique growing behavior and have risen (and continue to rise) in popularity with the younger generation of apartment dwellers and "urban gardeners." Because they don't require soil, *Tillandsia* appeal to the creative nature of millennials, the "maker" generation. A generation that values the unique and one-of-a-kind, this younger set of air plant enthusiasts find joy in the endless creative ways these plants can be displayed.

While most *Tillandsia* species are easy to grow, they do require some care and a proper environment in order to thrive. In the following sections, we will cover all aspects of caring for and growing your *Tillandsia*.

Determining Air Plant Care

The most common misconception about air plants is that they don't require any care at all. While *Tillandsia* are much lower maintenance than many types of plants, they do require some care and proper environmental considerations in order to thrive. Most air plant species originate

This *Tillandsia funckiana* still has its roots attached, which can be helpful for anchoring the plant into a wreath, frame, or other display. The root system on air plants does not take in water or nutrients and is used by the plants in their natural habitat to attach to a host—such as a tree limb or rock—where it will position itself best to maintain ideal light conditions. Air plants do not need their roots in order to survive and, when kept as houseplants, you can trim the roots of the plant to "clean up" the look, or use their roots to help attach them to your favorite display.

from tropical climates, and while they can certainly thrive as houseplants kept indoors in cooler or drier areas, their care should be adjusted depending on your environment.

Air plants are epiphytes, meaning that in their natural environment, they tend to grown on another tree, host, or object. The "roots" that you see on most *Tillandsia* are not actually root systems to absorb nutrients, but rather are used by the plant to anchor itself to its host object.

A tray of air plants sit outside on a patio table where they bask in the dappled, filtered afternoon sunlight. Air plants can do well outside in temperate climates as long as they get the right amount of light and humidity.

Air plants do not steal nutrients from their host, however, and only use their root system to anchor to their host as a home on which to grow. Air plants absorb nutrients through their leaves, using tiny hair-like vessels called trichomes to capture nutrients and moisture from the air in their environment.

The major factors when it comes to *Tillandsia* health are: air, light, and water.

While both mesic and xeric types of air plants share common factors, their differences will determine the different levels of light and water that they prefer. The major differentiating factor between the two types of air plants is the amount of trichomes they display; trichomes greatly affect both the look of the plant and how much light and water it will want in order to thrive.

TRICHOMES

It is through the trichomes—the fuzzy white hair-like things on their leaves—that air plants take in nutrients from the air and water. Trichomes are made up of many little cells, some living and some that are dead. When water comes into contact with the dead cells of the trichome, it swells them like a sponge and is absorbed into the living cells, which are then able to use the water to hydrate and nourish the plant.

The *Tillandsia tectorum Ecuador* gets its fuzzy white appearance from its extreme abundance of trichomes on its leaves. This air plant actually prefers bright light and even enjoys full, direct sunlight for periods of time. It requires very little water, and in fact can be susceptible to overwatering. Because of this, they should never be soaked but instead will do well with periodic mistings as its sole source of humidity.

Trichomes, which are most abundantly found on air plants that are native to desert-like conditions, also serve the purpose of holding water in once it is absorbed. The reflective nature of the trichome allows the plant to both retain more moisture and shield the leaves from the harsh rays of the desert sun in their natural habitat. Air plants that have an abundance of trichomes are often more drought-tolerant, but also prefer not to be over-watered, and can in fact be more harmed by too much humidity than by a lack of it. Trichomes also help shield the plant from direct sunlight, which allows plants with a lot of trichomes to happily grow in brighter environments.

The *Tillandsia xerographica* is a xeric species. Also known as the "Queen of the Air Plants," it has wide, silver leaves and prefers less frequent watering and bright light. In bright direct light, this sliver-toned plant will blush a red tinge.

Hundreds of *Tillandsia bulbosa Guatemala* cling to a wire wall in the partially-shaded light of a nursery. These mesic *Tillandsia* are native to moderately humid environments where they receive indirect sunlight under the protection of the South American forest canopy.

One of the most vivid examples of trichomes on a *Tillandsia* can be found on the *T. tectorum Ecuador*. These fuzzy white air plants get their distinctive "snowbwall" look from their many prominent trichomes.

Because of their abundant trichomes, the *T. tectorum Ecuador* actually prefers mistings to the full soaks normally recommended for most *Tillandsia* species. They also like more bright light and are one of the few *Tillandsia* varieties that can handle direct sunlight.

Xeric *Tillandsia*

The term "xeric" is from the Greek word for "dry," and xeric *Tillandsia*, or xerophytes, can thrive in drier climates. They are characterized by their abundance of trichomes, which give these plants their silver appearance and allow them to make the most of minimal moisture. These plants generally prefer less frequent watering and can tolerate (and even prefer) more direct sunlight than other varieties. Xeric air plants generally have silver leaves and are much stiffer than their mesic relatives.

Mesic *Tillandsia*

The term "mesic" is from the Greek word *mesos*, which means "middle." A mesic habitat is characterized as one that has moderate moisture, and this is the category that many *Tillandsia* species fall under. Mesic *Tillandsia* originate from moderately humid habitats like South American forests, and these plants prefer moderate humidity and frequent watering.

They will often have medium to dark green leaves, with less abundance of trichomes, and as such prefer moderate moisture as well as indirect, filtered sunlight. These plants grow naturally in forest habitats where the sun is filtered by the tree canopy. Some mesic varieties, like the *brachycaulos*, can actually blush red when given slightly higher levels of indirect sunlight. However, too much light can harm mesic varieties, and they may suffer from sunburn and develop brown spots.

A common distinction for many *Tillandsia* is actually "semi-mesic." These varieties display green leaves but have a higher proliferation of trichomes than their fully mesic counterparts. Many *Tillandsia* fall into this category, and these plants will prefer semi-frequent watering and a significant amount of bright indirect (not full) sunlight.

Adaptation

Like many plants, *Tillandsia* have the keen ability to adjust their characteristics and features—both physical and functional—to their growing environment. This is especially true for semi-mesic air plants, which can adjust to be either more or less mesic depending on their environmental factors.

For example, a species that is grown in a more consistently humid environment will naturally develop less trichomes and have a greener appearance, and produce offspring like it. At the same time, the same species of *Tillandsia* grown in a less humid environment that gets more sunlight will develop more trichomes, which can affect the color of the leaves. The shape of the leaves and even of the plant as a whole can change a bit depending on the environment that it's grown in, and this will influence generation after generation of *Tillandsia*. This is why there can be significant difference in appearance between two plants of the same species from different areas of the US, by different growers.

Air, Water, and Light

There are three main factors you need to be aware of when determining the right care for your *Tillandsia*: air, water, and light. The amount of light and water that the plants prefer will vary depending on many factors within the plants' environments. As a general rule, the amount of light that a *Tillandsia* will require will be proportionate to the amount of moisture it receives, and vice versa. A *Tillandsia* that gets more sunlight will also need more water as it will dry more quickly.

Temperature also plays into this sliding scale. A warmer, drier climate may require you to not only water your air plants more, but also to give them less sunlight.

Most *Tillandsia* purchased by customers come from a greenhouse with high light levels and plenty of moisture, so keep in mind when you first bring home a new plant that it may take some time to adjust to its new environment. Eventually the air plant will begin to acclimate to its new environment, which will allow it to thrive in its new conditions. Below we break down the three most important aspects of *Tillandsia* care and go into more detail about what makes air plants happy (and what makes them decidedly unhappy).

<< A stunning specimen, this *Tillandsia brachycaulos* hybrid is blooming, showcasing a number of blushing offsets. This mesic *Tillandsia* will do well with a good amount of indirect, filtered light, and with a bit more light will blush the red color you see here. This plant will not like direct sunlight and will appreciate more water with the more light that it receives.

LIGHT

In general, if you are keeping your *Tillandsia* indoors, you will want to make sure that they are near an adequate light source that gives a good amount of indirect light. This could be within 3–5 feet of a window, or near a bright artificial light source.

As mentioned earlier, the type of air plant will greatly determine how much light it prefers. For most mesic varieties, it is important to ensure that they don't receive too much direct sunlight. This is especially true when the plants are indoors, but also applies to plants kept outdoors. If you live in a temperate climate and you are keeping your plants outside, make sure they are in a shaded area that does not receive full sun. During the winter months, early morning sun or late afternoon sun is OK for short periods of time.

There are only a few varieties that can handle direct sun, including the *xerographica* and *tectorum Ecuador*. These plants, of the xeric variety, are much the exception to the air plant light rule as they actually thrive in direct sunlight. The *xerographica* will actually start to blush a red color when it receives high levels of sunlight.

Most mesic air plants love being placed in a bathroom or kitchen window that receives good indirect light, as the moisture from steam or sink spray will keep them hydrated and can result in them needing less frequent watering.

Bright artificial light can often be a sufficient source as well for mesic varieties, but in this case it's important that the plants are kept in closer proximity to the light source and that it is illuminated for the majority of the day. If you keep your plants indoors, it's sometimes beneficial to rotate them or move them outside once in awhile to ensure they are getting enough light.

WATER

The amount of water and frequency of watering, as well as watering methods, should again be adjusted by plant type as well as environmental factors. Mesic plants, which cover the majority of air plant species, will generally be happiest if watered at least once per week if kept indoors. Xeric varieties can go longer without watering, and the exceptional *tectorum Ecuador* plant only likes occasional mists or a quick dunk in water.

The *Tillandsia xerographica* loves sunlight and will thrive in higher levels of light than mesic or semi-mesic air plants. Here the *xerographica* soaks up some late afternoon sun on a patio table in Florida.

Three air plants soak in a bowl of rainwater, collected and stored outside for weekly watering. If it's not possible for you to collect rainwater or pond water, tap water that has been let to sit for a few hours prior to soaking is a good option. Just make sure the chemicals have time to dissipate from the tap water as they can be harmful after repeated exposure.

Indoor air quality can also determine the frequency of which air plans should be watered. Plants that are constantly being dried out from air conditioning or heaters, or which are in drier climates like Arizona, will require more moisture than a plant that is located in a more humid environment. A plant in a really dry climate could need daily soakings while one in a more humid environment might only need to be soaked once every 10 to 14 days or so.

A general rule of thumb is to give your *Tillandsia* a quick soak once or twice a week; if you live in a dryer climate, misting in between is very beneficial. The important factor is that the plants are able to dry out within four hours or so of watering.

After the plants have soaked for a bit, gently shake any excess water from their leaves. Then place air plants face down on a towel to dry so that any excess water drains out of leaves and pockets. If your plant is blooming, however, you should avoid placing the plant on the flower as this could crush it.

There are several ways to water your *Tillandsia*, but the most popular methods are soaking, dunking, and misting.

Soaking

Most air plants should get soaked about once a week, again depending on environment and plant type.

To soak your air plants, place the plants face down in a bowl, sink, or container and let soak for 10 to 20 minutes. After their soak, it is best to gently shake any excess water off the base of the plants, or any water that might be collecting in leaf curls or cups, as water that sits in the leaves can cause rot and damage or even kill the plant.

Take care when watering blooming plants not to soak the bloom flower, as it could cause the flower to fall off. Here, we carefully dip the sides of this blooming *ionantha* in water, taking care not to get the delicate purple flower wet.

After gently shaking any excess water from their leaves, lay the plants face down on a towel or cloth so excess water can drain and allow them to properly dry before replacing in a container. Putting them in front of a fan or somewhere with really good air circulation is highly advised, and will help reduce the chances of the plant developing rot or a fungus from staying wet too long.

If you have neglected to water your air plant in some time and it appears to be "thirsty," you can leave it to soak for a few hours, or even overnight. Many air plants will curl their leaves a bit when dehydrated, indicating that they require more water. Try to minimize the longer soaks to 12 hours or less.

Dunking

You can also water your plants using the "dunking" method. Dunk the plants several times and then gently shake off excess water. You can also let the plant sit for a minute or so to absorb some water. This can be good to do for plants like the *xerographica* that don't need as much hydration or as frequent soaking.

Misting

You can also mist your plants with water as needed in between soaks, but misting should not be used as a sole source of hydration for most plants. Again, the exception here is the fuzzy *tectorum Ecuador*, which does not like soaks and only should be misted with water. Particularly in dryer/warmer climates, the misting can help prevent the leaf tips from getting too brown. The whole plant should be misted enough that it's fully drenched in water.

Watering Blooming Plants

Few things in life are as exciting as when your well-cared-for air plant finally blooms! For blooming plants, though, special care should be taken when watering them. You should not dunk or soak the bloom flower, as this can cause the flower to wilt and rot. Instead, run the rest of the plant under slow-running water to get it wet, or dunk/mist only the portions of the plant that allow you to avoid wetting the bloom.

When to Water

It is best to water air plants in the morning, as this will allow them to properly dry. Air plants take in their nutrients through their leaves in the dusk or evening hours, and if the leaves are wet, this can block their ability to do so. Therefore, a morning soak will give the plant plenty of time to dry and properly put to use that hydration.

We recommend that you water the plants in the morning, and leave them out of their containers in an area that they can dry within four hours. Never let your air plant sit in water for prolonged periods of time, and always make sure they are dry before putting them back in a terrarium, shell, or other display.

Water Type

For the most part, air plants are not too sensitive to water condition. However, over time they can be negatively affected. The best type of water to use, whenever possible, is rainwater, pond water, or even aquarium water, as they have the good nutrients that your air plants will love.

You can also use tap water, but let it stand for several hours prior to watering to allow for any chemicals to dissipate that might be in the water treatment, such as chlorine, which can damage the plants with regular use. Bottled water and spring water are also fine to use, but never use distilled or artificially softened water! The salts in the water can rob the plant of vital nutrients and over time it will decline in health.

Over-watering can cause your air plants to rot, so don't assume that more is better when it comes to watering!

AIR QUALITY

Not surprisingly given their name, air quality is critical to the health of your air plants. *Tillandsia* need good clean air circulation to survive and thrive, and the quality of air can have a big effect on how much water the plants should get as well.

Humidity: most air plants originate from temperate climates with moderate humidity levels. With the exception of a select few xeric species of *Tillandsia*, most air plants prefer a level of moderate humidity, and don't do well in extremely dry conditions. While they can be kept indoors in places that are drier in climate, they will likely need more frequent soaks, with misting in between, to compensate for the lack of humidity. If they are kept in a more humid climate, they may need less frequent watering.

Air plants kept indoors should not be stored near an air conditioner, heat vent, or radiator, as this can cause the plants to dry out.

Air Circulation: Air circulation is critical to air plant health. It is important that after watering, the plants have enough air circulation to dry within 4 hours. This also means that any container that you use to display your plants should be at least partially open to ensure good air circulation and that it won't trap too much moisture. *Tillandsia* should not be displayed in enclosed containers, and it's important to ensure that they are completely dry before they are put back into a container that might restrict air circulation, as trapped moisture in the plant or against it could cause rot.

Temperature

Another critical factor to air plant health is the air temperature. Air plants can be found living naturally in a range of temperatures from 50 to 90 degrees Fahrenheit (10 to 32 degrees C). Air plants are, for the most part, tropical plants, and while they can live happily indoors, they should be kept in an environment where the air temperatures are in this range. Air plants can very easily be damaged by the cold, so it's best to ensure that they are not exposed to freezing temperatures. If you are keeping your plants indoors in a colder climate, don't let the plants sit against or even too close to a cold window.

An air plant can often adapt to its growing conditions (the level of light, humidity, and air temperature), but a drastic change in its environment can shock the plant and cause it to perish.

Fertilizing Your Air Plants

Fertilizing your *Tillandsia* can help them thrive. Feeding your plants will help ensure that they receive adequate levels of the nutrients needed in order to have healthy growth, new blooms/flowers, and overall well-being.

A lower nitrogen-based fertilizer, when used properly and sparingly, can help promote growth, pup (or off-set) production, and can encourage the *Tillandsia* bloom cycle. There are several fertilizers on the market that are formulated specifically for *Tillandsia*, but you can also use a diluted orchid or bromeliad fertilizer. It's important to use a non-urea nitrogen fertilizer, as this type of nitrogen needs bacteria in the soil to break down into usable nitrogen for the plant. Since air plants absorb the nutrients through their leaves, they are unable to thoroughly process that type of nitrogen and it could ultimately damage the plant.

Fertilizers are described with a NPK rating, giving weights of the amounts of nitrogen, phosphorus, and potassium within the mixture. In order for *Tillandsia* to survive, or any plant for that matter, they need these three elements in their life. Having sufficient levels on nitrogen will help the plants with new leaf growth and leaf length. Nitrogen also helps with tissue repair in the event there is leaf damage. Phosphorus will help encourage the blooming and flowering process, while potassium is needed for overall plant growth and to ensure the plants' core functions are operating in a healthy manner.

Fertilizer should be used sparingly, as air plants can be very sensitive to over-fertilization. A good rule of thumb is to mist or soak the plants with a small amount of diluted fertilizer once per month.

This small grouping of air plants sits on dry river rock in an open glass container that allows for plenty of air circulation.

Trimming/Pruning Air Plants

It is natural for air plants to lose some leaves as new ones are produced. Some varieties (particularly those with more "wispy" leaves) can experience a slight browning of the leaf tips. Dead leaves can often be gently removed without harming the plant, and many people will opt to trim the tips of the leaves.

You may also prefer to trim the roots of the plants for aesthetic purposes, and this is absolutely fine to do. The roots are solely used to anchor the plant to its host, and aren't required for nutrient intake. It just depends what you wish to do with the plant—sometimes those roots can be useful for displaying your air plant if you want to use them to help anchor the plant (for example, on a wreath or frame). The root system will grow back over time.

Common *Tillandsia* Health Issues

Pests and Ailments

Most *Tillandsia* are very resilient to pests, but there are several types of pests that can affect your *Tillandsia*. Luckily most of them are very easy to treat. Common pests that affect *Tillandsia* include scale, such as mealy bugs, and aphids.

To treat for these pests, you can often spray the bugs out of the plant with water, or use a non-toxic solution of dish soap. Another common and effective treatment uses rubbing alcohol; simply soak the tip of a cotton swab and dab the saturated swab on the leaves to repel pests. For larger infestations, mix regular dish soap or baby shampoo with water and fully coat the plant in this solution. After a few minutes, make sure to fully rinse the plant of any soap. Often these steps will be sufficient to ameliorate the problem, but if pests remain after this method, an insecticide will need to be used.

SCALE

Scale insects are small insects that create a waxy barrier to protect themselves while sucking the sap from a plant. There are many types of scale insects, and they vary widely in appearance. The most common scale insect that affects *Tillandsia* are mealy bugs.

Mealy bugs are small scale insects that feed on the sap from the plant, and secrete a waxy, cotton-like substance that protects them from predators. If your *Tillandsia* leaves seem to appear covered in a waxy white powder that resembles little tufts of cotton, that is a pretty sure sign of a mealy bug infestation. To rid your plants of mealy bugs, you should first and foremost quarantine the plant from any other plants, and then use a natural dish soap on the plant. The soap can be dabbed on the leaves with cotton swabs, or diluted with water and sprayed on the plant. The dish soap works by creating a barrier on the plant that will smother the pests. Because of this, it's critically important to make sure that any soap residue is thoroughly rinsed off the plant after treatment, as the soap, if left on the leaves, could cause the plant itself to suffocate.

Aphids

Aphids also invade by sucking the sap from a plant, but it is rare that they will do significant damage to the plant, unless the infestation is particularly severe. Aphids are tiny bugs that may be red, green, yellow, or black. Most of the time aphids are found hiding out in the shelter of a leaf at the base of the plant. Aphids are normally easy to treat for, and often can be knocked off the plant using a stream of water. If this doesn't remove them all, a diluted soap spray as previously recommended can be used.

Rot

While proper hydration is important to *Tillandsia* health, rot due to excessive moisture or prolonged saturation is one of the most common issues that can lead to the death of these plants. Though *Tillandsia* are relatively resilient, too much moisture can make them rot from the inside, and when this happens, the plant will simply fall apart. If your *Tillandsia* displays a black and mushy base, starts to fall apart, the interior of the stem appears black and soft, or the plant displays mold, these are all signs that it has been affected by rot.

Rot often occurs as the result of water having been let to stand in the leaves of the plant, or from keeping the *Tillandsia* on a moist surface or in an environment where it has not been allowed to fully try. To avoid rot with *Tillandsia* kept indoors, it is important to remove the plant from any display, shell, vase, or terrarium to water it, and after watering, the plant should be in an area that allows it to completely dry within 4 hours. Never water your plants in their container, as moisture that sits in the terrarium or display can cause the plant to rot. Whenever your plants get wet (be it from rain, if outdoors, or from manual watering), it's a good idea to gently shake the excess water out of the plant and put it in an area with good air circulation. If you are in a more humid environment, the use of a fan can help accelerate the drying process.

Unfortunately, once rot has started in a *Tillandsia*, it is difficult to curb and will likely destroy the plant, so the best medicine in this case truly is prevention.

Sunburn

While *Tillandsia* do love very bright indirect light, prolonged periods of direct sunlight can damage the plants. Sun damage can show up in several forms, depending on the plant type and the extent of the damage. Signs of sun damage include a browning of the plant tips, brown or yellow spotting, or in more extreme cases, the plant can completely dry up and turn brown. There are few varieties that can handle full sun, and most *Tillandsia* should be kept in a shady area that receives only filtered or indirect light.

Cold

Tillandsia can also be affected by cold weather. While some varieties can handle lower temperatures than others, they all will suffer if left in freezing conditions. If your air plant is harmed from cold weather or freeze damage, the plant will normally turn a dark brown/black color and become mushy. Unfortunately, a plant that has suffered any type of cold damage can rarely be saved and will continue to decline until completely dead.

Tillandsia Varieties

There are an estimated 700 different varieties of *Tillandsia*, with more still being discovered. However, not all *Tillandsia* varieties are easy to grow, and therefore only a limited number of varieties are sold commercially. There are several species of *Tillandsia* that are protected by CITES (Convention on International Trade of Endangered Species of Flora and Fauna) and prohibited from being collected or harvested from their natural environment. Because of these restrictions, commercial *Tillandsia* growers have continued to expand their collections by developing a large number of hybrid species over the years, many of which have become nearly as popular as some original *Tillandsia* species. *Tillandsia* hybrids are also found in nature as plants cross-pollinate.

Tillandsia (as well as many plants in the *Bromeliad* genus) species can have many varieties within the species, and while these varieties share some common traits, they can have distinctly

different looks from other varieties in the same species. The different traits that these varieties display is due primarily to the differences in their originating climates—how and where they are grown. *Tillandsia*, like many plants, will over time adapt to the environment in which they are grown, and these adaptations can gradually change the form of the plant as it breeds new generations. For example, a *Tillandsia ionantha Mexican* and a *Tillandsia ionantha Guatemala* will show different traits due to their country of origin.

While there are too many varieties to list them all in this book, in the coming pages we will discuss some of the more common and well-known *Tillandsia* varieties, as well as some of the rarer forms and hybrids.

Common Commercial *Tillandsia* Varieties

TILLANDSIA IONANTHA

The *Tillandsia ionantha* is one of the most common and widely-purchased air plant species. This species comes in many different varieties, with characteristics that are determined in large part by their geographic location of origin.

One of the most widely available *ionantha* varieties is the *ionantha Guatemala*. A larger *ionantha* variety, a single *Guatemala* can grow to be as large as 5" tall. Other common *ionantha* varieties include the *ionantha Mexican, ionantha rubra, ionantha scaposa (kolbii), ionantha fuego,* and *ionantha peanut*. Rarer varieties include the *ionantha peach, ionantha druid,* and *ionantha zebrina*.

The *ionantha* variety is generally a fast grower that is prone to grow in clumps forming a ball over time if left together.

Ionantha Guatemala

Origin: Native to Guatemala, with an average size of about 1/2" at the base, 2" in width, and 1 to 5" in height. Its leaves are spikey and thin in form and covered in trichomes, giving this air plant its silvery-green color, which transitions to a light pink/red hue throughout its bloom cycle. This plant displays a beautiful contrast in color when in bloom, sprouting stunning purple spike sprouts from its center, which emits a delicate yellow flower.

Ionantha Mexican

Originating from Mexico, this *ionantha* variety will grow to an average size of 1 to 3". With high levels of sunlight, it will blush beautiful shades of red, yellow, and orange. Like most other *ionantha* forms, this plant blooms a delicate purple and white spike with yellow flowers. After the bloom cycle, the plant will produce offset, and eventually form a clump.

The *Tillandsia ionantha Mexican* has many spikey leaves. This mature specimen has grown vertically, a contrast to younger versions of this species, which can take on a rounder appearance.

Ionantha Rubra

Native to Central America, there are two different types of *ionantha rubra*: "soft" and "hard." The soft form of *rubra* generally grows shorter and wider, with soft leaves that will blush a beautiful red color when given high levels of sunlight. The "hard" variety tends to grow taller and more vertically, with brighter green leaves, and when blushing shows bright pinkish-red tips. Both forms bloom like other *ionantha* varieties and produce pups after flowering.

Ionantha Fuego

Also native to Central America, this smaller *ionantha* variety is known for its bright red color, which can be maintained year round with adequate levels of light. A unique characteristic of the *fuego* is its significant change in shape throughout its growth process. Starting small as a short and wide plant, the leaves will tend to grow more upright as it matures, drastically changing the shape over time as it changes size. This form of *ionantha* will also pup after flowering, and can easily be grown into a ball or "clump."

The *ionanthan fuego*, true to its fiery name, is known for its brilliant red hue. These *Tillandsia* have become some of the more rare *ionantha* species, due in large part to their coveted nature.

TILLANDSIA AERANTHOS

This *Tillandisa* species is native to Central America, and is a rather hardy plant in the right environment. Its leaves, stiff and slightly sharp to the touch, grow upward from the center in an almost cone-like fashion. Normally a beautiful, healthy green hue, the slender funnel-shaped leaves will blush as they mature. When in bloom, it emits a vibrant delicate flower in pink, blue,

The *Tillandsia aeranthos* is coveted for its striking bright pink bloom.

and purple tones, and this bloom can last several weeks. There are several hybrid versions of the *aeranthos* developed commercially because of its colorful blooms and symmetric appearance.

TILLANDSIA BRACHYCAULOS

The *Tillandsia brachycaulos* is a green-leafed species native to Central America, including Mexico and Guatemala. It is semi-mesic, meaning that it prefers indirect light and higher levels of moisture. This is one of the few mesic varieties that can handle higher levels of light, and will

The *Tillandsia brachycaulos* has bright green leaves that grow in a circular fashion.

The *brachycaulos x concolor* hybrid *Tillandsia* blushes red when blooming, with a red flower.

The *Tillandsia brachycaulos x abdita* is a popular hybrid for its brilliant red color. This air plant's natural color is often commercially enhanced with a red floral tint by nurseries, as is shown in this photo.

blush a deep red when exposed to brighter light conditions. The *brachycaulos* can emit multiple blooms from the center of the plant, which produce a white and yellow flower. The flowers on these blooms, however, tend to fade quickly.

There are also several hybrid species of the *brachycaulos*, among them the *brachycaulos x ionantha*, *brachycaulos x concolor*, *brachycaulos select*, and *brachycaulos x streptophylla*.

TILLANDSIA BERGERI

The *Tillandsia bergeri* is a fast-growing species that is very similar in appearance to *Tillandsia aeranthos* (with the exception of its bloom stalk and flower). This type of air plant is very prone to clumping and, unlike most other *Tillandsia*, which have limited propagating windows, will continue to produce offsets year round. This enables the clump to get rather large pretty quickly. When blooming, the *bergeri* produces beautiful light blue and white flowers.

TILLANDSIA BULBOSA

This bulbous air plant species has two different varieties: the *bulbosa Guatemalan* and *bulbosa Belize*. The *Guatemalan* form is smaller and generally darker, while the *Belize* form grows much larger and generally has a brighter, more colorful appearance.

Both *bulbosa* varieties will emit a nice bloom tract with violet flowers. They will both also create pups around the base of the plant, and over time will grow to form a large multi-plant clump.

Tillandsia Butzii

Native to Central and South America, this bulbous *Tillandsia* is dark green in color and dappled with spots in darker

The *Tillandsia bulbosa* gets its name from its bulbous shape.

Tillandsia bulbosas will form a clump if pups are not separated. This healthy clump of *bulbosas* will continue to grow more pups into a larger and larger clump.

shades of green, black, and even dark purple. The slender leaves of the *butzii* grow upwards and eventually will cascade down, giving this variety a decidedly aquatic look. When blooming, the *butzii* will send out a tall bloom spike that emits violet flowers. Over time, the *butzii* will continue to produce pups and eventually form a nice ball of plants, which can be very hardy and require minimal care in the right environment.

TILLANDSIA CAPITATA

This large-growing *Tillandsia* is found in Mexico, Cuba, and some areas of Central America including Guatemala. There are several different variations of the *capitata* that include *capitata peach*, *capitata yellow*, and *capitata maroon*, among others. The leaves tend to grow in a very elongated

The beautiful *Tillandsia capitata* is a personal favorite of ours for its full, beautiful shape.

fashion, and will blush when starting the bloom process. Like the *brachycaulos*, the *capitata* will emit several purple bloom tracts that will display delicate white and yellow flowers. This type of *Tillandsia* prefers adequate moisture and indirect light.

TILLANDSIA CAPUT-MEDUSAE

The *Tillandsia caput-medusae* is sometimes referred to as "Medusa's Head," which is fitting given the plant's serpent-like leaves. Generally bright green in color, this type of *Tillandsia* can also display shades of purple and will sometimes develop a silver appearance when placed in a very bright environment. When in bloom, it will emit a long bloom tract that extends above the plant and can be very colorful when given bright light.

Don't stare too long at this crazy plant or you'll turn to stone! Just kidding; go ahead and stare all day at this beautiful *Tillandsia caput-medusae*.

TILLANDSIA CONCOLOR

Native to areas of Mexico and Central America, this stiff-leafed variety is known for its bold colors and amazing inflorescence shown in shades of bright yellow, red, and green. This species of *Tillandsia* prefers bright light, and will blush shades of red with sufficient light.

TILLANDSIA FUCHSII (COMMON NAME: ARGENTEA)

Native to Mexico, Cuba, Guatemala, and Jamaica, this unique *Tillandsia* is characterized by its small, compact base, from which many slender, pale, silvery-green leaves extend out in every

The *Tillandsia fuchsii*, with its bulbous base and wispy leaves that extend in a symmetrical, spherical shape, is striking enough on its own without any other plants. Here it hangs in a simple glass globe on a semi-shaded patio, enjoying filtered sunlight and temperate Florida air.

direction, like wispy tentacles. It averages in size from 2 to 4" in height and 1 to 3" in width. As this plant matures, it will release a long and slender shoot straight out of the middle that blooms a beautiful flower from the tip. It will normally go through its bloom cycle in late spring, and it produces one of the most unique blooms of all *Tillandsia* species.

TILLANDSIA FUNCKIANA

Appropriately named for its funky appearance, this *Tillandsia* twists and turns in all directions throughout its growth cycle. Grassy green in color with a small woody base and spiky leaves, it develops tiny pups that turn into larger clumps. As it starts its bloom cycle, it will turn bright red and soon after emit a beautiful, bright red "lipstick" bloom with yellow flowers.

TILLANDSIA HARRISII

This air plant is named in honor of Bill Harris, an avid air plant enthusiast who was murdered in Guatemala in 1985. This plant is listed on Appendix II of the international CITES (Convention

The funky *Tillandsia funckiana* blooms a bright lipstick-red flower.

on International Trade in Endangered Species) agreement, meaning that trade in this species must be accompanied by a Federal Form A certificate that shows the plant was produced artificially and not collected from nature. Their average size ranges from 1" at the base, 2 to 3" wide, and 4 to 5" tall, although it's been known to reach up to 8" in height! This hardy plant features soft, succulent, silver/grey leaves with an abundance of trichomes. The elongated leaves are symmetrical and grow around its long stem. These special *Tillandsia* will produce a long-lasting, bright red inflorescence with red and violet flowers when in bloom, a lovely contrast to its light-colored leaves.

TILLANDSIA XEROGRAPHICA

Known as the "Queen of Air Plants," this is a very slow-growing epiphyte that can reach an impressive size. Its name (pronounced "zerografika") is derived from the Greek words ξηρός (*xeros*),

A birds-eye view of the *Tillandsia xerographica* shows its unique and impressive shape, its leaves curving in a giant sphere. It is easy to see why this is such a popular and coveted air plant.

meaning "dry," and γραφία (*graphia*), meaning "writing," due to its elegant, artful leaves. Native to southern Mexico, El Salvador, Guatemala, and Honduras, this unique xeric *Tillandsia* is one of the most popular air plants among air plant collectors. Although they are now widely available, at one time their popularity almost led these beautiful plants to extinction. Because of this, this is also a species protected by CITES, and cannot be harvested from nature.

These round, cushion-y *Tillandsia* can range greatly in size and can grow quite large over time. Small sizes ranges from 3 to 4" in height and 4 to 5" in width, and larger sizes range from 5 to 6" in height and 8 to 10" in width. Size and shape can also vary depending on how hydrated

This "Queen of the Air Plants" is literally put on a pedestal by a sunny window. These stately *Tillandsia* prefer higher levels of sunlight than most *Tillandsia,* and can go longer between waterings.

the plant is; drier plants will have more tightly-curled leaves, while a more hydrated plant can be "teased" out to have looser leaves. Silvery-green in color with thick, wide, curly leaves, this hardy plant is one of the few *Tillandsia* that can handle (and actually prefer) full sunlight. With consistent bright light, they can blush a reddish color. While slow to bloom, during its bloom cycle leaves will blush a light pink and a tall, spiky growth will sprout from the center adding a contrast in form and sometimes emitting a purple-toned bloom.

This plant is drought tolerant and can survive in desert-like conditions, and because of this it actually should be watered less than most *Tillandsia* species. It can be misted as its primary hydration source, with a soak every few weeks. It's important that after soaking these plants, you gently shake the excess water from the leaves by turning it upside-down, as this plant can easily collect and retain water in its wide leaf base.

This stunning plant can be displayed in a variety of ways, and due to its size is often displayed on its own. Due to its stiff curling leaves, it's a great option to be mounted on a frame. This *Tillandsia* is also perfect as part of a floral bouquet, and can even be used on its own as an alternative bridal bouquet.

TILLANDSIA TENUIFOLIA

The *tenuifolia* is native to Central America and South America. It's a thick and bushy air plant variety with stiff, sturdy leaves that grow in a natural curve as it matures. *Tenuifolia's* grassy green leaves stem from a soft yellow base and often turn to a unique bronze color as it ages. During its bloom cycle, the beautiful pink inflorescence and

This *Tillandsia tenuifolia* sprouts a pup from its tip, unlike many other *Tillandsia* species that produce pups from their base.

This simple but striking bouquet puts the *Tillandsia xerographica* in the spotlight, shown here with simple floral accents. >>

royal blue to violet colored flowers will attract butterflies, bees, and birds if placed outside. There are several varieties of this species including *Emerald Forest*, *Bronze*, and *Black*.

PSEUDOBAILEYI

The *pseudobaileyi* is one of the more popular nature-collected species and is adaptable to many different environments. Within the *pseudobaileyi* species, there are two recognized varieties: Subsp. *pseudobaileyi* (found in Central America in Mexico from Chiapas north to Veracruz Nayarit) and Subsp. *Yucatanensis I. Ramirez* (found in the Yucatán Peninsula). One of most unique air plants in appearance, it has a tall, slender shape with twisting, cylindrical, tentacle-like arms. Its leaves are silvery-green in color and will display vertical maroon lines in the margins, similar to the brown lines on the skin of an onion. Its inflorescence also features a maroon-like color and eventually produces a purple flower with a yellow tip pollen at the end, the stem of which will stay prominent for several weeks to months. After the bloom dies, it will produce pups, which can remain attached to the plant to create a beautiful cluster or can be gently removed once the pup is one-third the size of the mother plant.

A hardy plant, this *Tillandsia* prefers light shade and a spray mist with an occasional soak, depending on its climate (it can be watered less often in a humid environment). Make sure it dries out completely before watering again to avoid rot.

TILLANDSIA VELUTINA

Known for its lush, soft, dark green leaves that curve out of its base in wide arcs, the *velutina* is easy to care for and thrives in bright, indirect or filtered light. As this air plant matures, its leaves will turn beautiful shades of yellow,

The *Tillandsia velutina* sits peacefully atop a piece of driftwood.

peach, and pink. When it comes to watering, you should mist it frequently and occasionally dunk under water, afterward allowing water to fully drain from the base.

TILLANDSIA STRICTA

The *Tillandsia stricta* is one of the fastest growing plants from seed, and there are many different types of cultivars and hybrids. Some of the more common cultivars include: *stricta soft*, *stricta hard*, *stricta silver*, and *stricta midnight*, among others. Generally dark green in color, most *strictas* have stiffer leaves and will bloom a nice white and pink bract that will open to display white flowers. A very popular *Tillandsia*, these are generally easy to grow and will thrive in a large range of conditions, creating a nice sized clump over time if left intact.

TILLANDSIA JUNCEA

Tall and grassy, the *juncea* is native to Mexico and Central and South America. This variety can get rather tall, and though mostly green and silver in color, it will turn shades of dark

The *Tillandsia stricta* has slender, dark green leaves and will bloom a beautiful pink and white flower.

The tall and slender *Tillandsia juncea* is coveted for its unique shape. This air plant will continue to grow taller and more full as it matures.

red and yellow when blooming. The bloom spike can take several months to develop before flowering.

TILLANDSIA TRICOLOR V MELANOCRATER

The tricolor is a stiff-leafed plant that grows in a rather upright position. It prefers bright, indirect light and humidity and will turn shades of dark red if given the right environment. The bloom spike gets rather large and displays three colors (hence its name): canary yellow, red, and dark blue. After blooming it can create up to twelve offsets, with some being created on a "stolon," which is another section of the plant base that grows off to the side.

The *Tillandsia tricolor v melanocrater* is unique for its stiff, dark base.

When in bloom, the *Tillandsia tricolor v melanocrater* will produce unique bloom spikes, here shown in shades of red.

TILLANDSIA USNEOIDES (SPANISH MOSS)

Commonly known as Spanish moss, this type of *Tillandsia* is very prevalent in the southern parts of the US as well as in Central America, South America, and even in the northeast of the US. Generally found hanging from the limbs of trees, this plant will clump together easily and can create a dramatic look over time as it grows quite large. Throughout history, Spanish moss was also used for many practical applications, including mattress stuffing, building insulation, landscape bedding, and even in the stuffing for car seats in the early 1900s!

Collector *Tillandsia* Varieties

There are many species of *Tillandsia* that are considered rare or collectible and the list is actually quite large. There are some rare varieties that can be sourced somewhat easily, and a few of the more common ones are listed below.

CACTICOLA

This rare variety is found naturally in northern Peru, and has wide, soft leaves. This air plant likes bright light and more temperate conditions. When blooming, the *cacticola* will emit a large inflorescence that is shades of purple in color and will last quite some time. Its delicate flowers are white with purple tips, making this plant a very nice variety for any collection.

CAPITATA DOMINGENSIS

A miniature form of the *capitata* that can be found in the Dominican Republic, this plant is known for its continual burgundy coloring. This smaller form is easy to grow in humid conditions with bright light.

CAPITATA GUZMANOIDES

This rare variety of *capitata* is a deep red color in appearance and will bloom in similar fashion to other bromeliads releasing a large bloom tract that will last for some time. This variety of *capitata* is rather hard to find, and grows in very few places naturally.

A single *Tillandsia edithae*.

EDITHAE

This silvery *Tillandsia* is one of very few air plants that produces

The rare *Tillandsia cacticola* produces a tall spike during its growth and bloom cycle. >>

coral-hued flowers when in bloom. Found on the cliff and rock faces of Bolivia, this variety enjoys bright, indirect sunlight and a lot of air movement. The *edithae* will create a good deal of offsets, which normally form around the base of the plant, evolving to a large clump over time.

TECTORUM

The *tectorum Ecuador* grows high in the mountains of Ecuador and Peru. Its appearance is fuzzy and silver, due to its abundance of trichomes, which help protect it from the sun as well as gather moisture and nutrients from the air. These xeric *Tillandsia* like lots of air movement, bright sunlight, and low humidity. These fuzzy snowball-looking plants should not be soaked, but rather just misted from time to time with water.

When the *edithae* clump, they do so end-on-end and can form an impressive, snaking series of air plants.

The fuzzy *tectorum Ecuador* gets its snowball-like look from its massive abundance of trichomes. These plants prefer bright light and low humidity and should only be misted—not soaked—with water.

TILLANDSIA STREPTOPHYLLA

Found in Mexico and parts of South America, this species of *Tillandsia* grows in a variety of elevation and forest conditions. Its appearance can be tight and curly if grown in drier conditions, and more open and wavy if given more moisture. These *Tillandsia* are normally green in color and will blush to a pinkish red color when exposed to higher levels of indirect light. The inflorescence is spectacular and can last for a long time.

The *Tillandsia streptophylla* has wide, noodle-like leaves. Here we see a large *streptophylla* with its roots intact.

Here we see a more tightly-curled version that has been grown with less water and light, affecting both its shape and color.

TILLANDSIA GARDNERI

Found in Central and South America, this silver-colored *Tillandsia* has soft leaves. It prefers higher levels of indirect sunlight and light humidity in a cooler environment. It will bloom a red flower bract with dark purple flowers.

The *Tillandsia gardneri's* soft, silvery leaves showcase an abundance of trichomes, leading these plants to prefer more light and less humidity.

TILLANDSIA SELERIANA

Found in Mexico to Central America, this larger, slow-growing *Tillandsia* is one of the largest bulbous species, and has a distinctly large, thick, bulbous base. Sometimes referred to as the ant plant, it's not uncommon for this *Tillandsia* to be colonized by ants in the wild. Its coloring is generally green, with visible silver trichomes depending on its light levels. When blooming, the tips will lighten in color as the plant begins to flower.

The unique *Tillandsia seleriana* has a large bulbous base, resembling a sort of root vegetable.

TILLANDSIA STRAMINEA

With its silver appearance, the *Tillandsia straminea* is found at higher elevations where the temperatures are cooler and the air movement is high. This *Tillandsia* has many different variations, with the most common being a thick leafed variety and a thin leafed variety, and the thick leaf variety tends to grow larger. The *straminea* will bloom purple, fragrant flowers, and over time tend to clump together if allowed to do so.

TILLANDSIA DURATII

Considered a "must have" for most *Tillandsia* collectors, this unique plant is also thought to be one of the hardiest species available, and is native to the dry forests in South America. Its velvet-like leaves grow upward to start, and then fall under their own weight to create a unique base of leaves that collect over time as new shoots continue to grow from the center

The *Tillandsia duratii* is unique for its long, winding shape and fragrant blooms.

The endangered *Tillandsia fasciculata* hybrid leaves grow in a vertical fashion. This plant will produce an impressive bloom when in its bloom cycle.

upwards. The *duratii* is also one of the few *Tillandsia* species that has fragrant flowers when in bloom. These delicate purple flowers have a grape-like scent and can last for some time. This specimen does best growing in a clump and will get rather large with many older plants reaching over 6' in length. Usually the *duratii* is grown hanging, where it can develop its natural shape and structure over time.

TILLANDSIA FASCICULATA

Native to Central America, South America, and some parts of the southern United States, this type of *Tillandsia* will grow rather large and will display a large inflorescence that is quite colorful. This plant is generally found in the wild attached to large oak or cypress trees in areas with higher levels of moisture and humidity. This *Tillandsia* is endangered in the US, but this is mainly due to the natural infestation of insects, rather than illegal harvesting or over-collection.

TILLANDSIA CYANEA

This type of *Tillandsia* is commonly referred to "Pink Quill," due to its large bloom spike that is bright pink in color and will last for a prolonged period of time. Native to the rainforest in Ecuador,

Rows of *Tillandsia cyanea* sit happily in a greenhouse, all showing their hot pink "quill" bloom spikes.

this is one of a few *Tillandsia* species that can be potted in soil. The *cyanea* should be potted in a fast-draining mix, as used with orchids. Though slow to do so, when the *cyanea* does finally bloom, it emits delicate pink and white flowers out of its dark pink bloom spike.

TILLANDSIA VERNICOSA

This species is found in Central American and some parts of northern South America. A hardy, stiff-leaved *Tillandsia*, they generally have dark green and grey coloring that can blush shades of purple. *Vernicosa* will develop an inflorescence of a very bright orange color, which can get quite

The impressive *Tillandsia vernicosa* produces one of my personal favorite blooms—a tall, salmon-pink spike with delicate white flowers.

large, and will produce delicate white flowers when blooming. Over time, this variety will easily create multiple offsets and clumps.

Some Popular *Tillandsia* Hybrids: *Aeranthos x Stricta*

BRACHYCAULOS X CONCOLOR HYBRID

This hybrid is from parent plants *brachycaulos* and *concolor*. Combining some of the best qualities of each plant, this hybrid has stiff leaves and is mostly greenish in color. With higher levels of indirect light, it will blush a bright red and then bloom with a dark red tract and purple and yellow flowers. This hybrid is a very hardy and colorful plant.

The blushing *brachycaulos* x *concolor* hybrid *Tillandsia*.

BRACHYCAULOS X STREPTOPHYLLA HYBRID

A hybrid of *brachycaulos* and *streptophylla*, this *Tillandsia* has green leaves that are soft and curly. With bright light, the leaves will blush a beautiful pink and red as it begins the bloom pro-

The *Tillandsia Eric Knobloch* is a hybrid of *brachycaulos* and *streptophylla*. When blooming, as shown here, it displays a vibrant pink and purple flower.

cess. Its hardiness can be credited to the *streptophylla*, and it can grow quite large in size. This hybrid is also commonly known as Eric Knobloch, named for its creator.

BRACHYCAULOS X IONANTHA

This graceful hybrid of the *ionantha* and *brachycaulos* can grow taller than 8 to 10", and with high levels of sunlight the new leaves will blush a beautiful pink and red. Due to the *brachycau-*

los traits, this plant has softer leaves and a larger appearance. The *ionantha* traits include nice color and the ability to handle less moisture and higher levels of light. This must-have hybrid is also commonly known as Victoriana.

CAPITATA X XEROGRAPHICA HYBRID

This is a hybrid of *Tillandsia capitata* and *Tillandsia xerographica*. With flowing, hardy leaves that are long and silvery-green in color, it is a stunning combination of the two parent plants. During its bloom cycle, its leaves will blush and produce a beautiful yellow bract with bright, bold purple flowers. This hybrid will produce pups more quickly than its parent *xerographica*, due to the *capitata*'s propensity to pup. It gets its hardy nature, though, from the *xerographica*, and can tolerate more sunlight and less waterings (depending on the environment it is grown in). Instead of the common air plant watering method of weekly soaks and mistings, give this air plant an occasional quick dunk in water and lightly shake off any excess to prevent rot.

The *brachycaulos ionantha* hybrid *Tillandsia*, also known as the Victoriana, takes on the best of both plants, with softer, wider leaves of the *brachycaulos*, but the heartiness of the *ionantha*.

The impressive *capitata x xerographica* displays clear traits of both mother species, taking its heartiness and size from the *xerographica*, but producing pups more readily like its *capitata* heritage.

FASCICULATA "TROPIFLORA"

A large growing natural variation of the *fasciculata*, this plant was originally discovered by growers visiting Jamaica. This type of *fasciculata* is thought to grow only in a few areas, and is rather rare. Growing to well over 30" at maturity, this *Tillandsia* will display a rather large inflorescence that is quite colorful and will last up to a year in the proper environment. This is definitely a rare collector's species!

STRICTA X RECURVIFOLIA "HOUSTON"

Cultivated in 1982, the *"Houston"* is a hybrid of the *Tillandsia stricta* and *Tillandsia recurvifolia*. It ranges in size from 4 to 6" tall and 3 to 4" wide. The stiff leaves on the *"Houston"* air plant

The silvery leaves of the *Tillandsia "Houston"* provide a beautiful contrast to its bright pink bloom.

<< The *fasciculata "tropiflora"* has an impressive bloom spike and can grow to be very large in size.

The *Tillandsia "Houston"* produces a bright pink flower when blooming.

grow from the base in a graceful curve and "glow" with iridescent silvery-green leaves, their unique appearance due to their abundant trichomes. At the start of the blooming process, its leaves will turn a beautiful purple color. Once in bloom, the *"Houston"* will emit stunning pink and white flowers that last up to a month. It will flower once again in two years and produce pups. One of the faster growing air plants, the *Tillandsia "Houston"* is prized for the clump of offsets that it can reproduce. Even within this hybrid species, the *"Houston"* has two distinct varieties—one which blooms a light pink flower, and another, known as *"Cotton Candy,"* which blooms a very rich, dark pink.

Xerographica x Fasciculata

This hybrid variety exhibits strong traits from both the *xerographica* and the *fasciculata*. It is very hardy, and will grow rather large. This *Tillandsia* will showcase a very large and impressive inflorescence that can last up to a year, and can be quite colorful when given high levels of indirect sunlight.

A close up of the bloom spike of the *fasciculata x xerographica* shows the dynamic range of colors it displays.

GROWING AIR PLANTS

There are two ways to grow *Tillandsia*: from seed and from propagation. Growing a *Tillandsia* from seed is a very long process as they are very slow-growing plants. It can be very rewarding though, as *Tillandsia* are more likely to be healthy and well-formed if given the right conditions throughout the growing process. Because the seed growth method is such a slow process, propagation is by far the fastest, easiest, and most prevalent method to grow these unique plants.

Growth Process

Like all living organisms, *Tillandsia* go through various distinct stages in their growth cycle. In general, plants that are grown from seed have a much longer growth process versus a plant that is propagated as an offset. As the plant matures, it will eventually bloom, marking the first stage of their reproductive process.

The *Tillansia seleriana* begins to spike, the first stage of the bloom process.

The wispy *Tillandsi fuchsii* will produce a slender vertical spike straight from its base.

The *Tillandsia capitata peach* will blush a pink color and emit flowers from its center.

The *stricta's* bud will then open to a large white and pink bloom flower.

There are many different "styles" of blooms within the *Tillandsia* world. Some plants, like the *capitata peach*, will emit the flowers straight from the center of the head of the plant.

Others like the *stricta* and *aeranthos* will have a small bud that eventually grows larger from the center of the plant and will open up to reveal its flowers. The flowers from these type of blooms tend to be short lived, with some lasting only a few days and others up to a few weeks.

Another much longer lasting type of bloom cycle can be found in larger plants like the *fasciculata x xerographica*, *caput-medusae*, *xerographica*, and several others. These plants will grow a large bloom tract called inflorescence that can be well over a foot tall! Over time, flowers will open up and release from the inflorescence. Some varieties will have a bloom tract that can last over a year.

After a *Tillandsia* has bloomed, the next stage of growth is the creation of offsets or "pups." The amount of pups created can vary depending on the variety as well as the conditions. For instance, a *xerographica* will usually only create one pup after flowering, so propagating this type of *Tillandsia* is very difficult. Other varieties like the *ionantha* will continue to create pups over time and will eventually form a clump. Some clumps that are very old can have hundreds of plants that create a giant ball/clump. There is something to be said for leaving your offsets un-separated and allowing them to clump, as over time when the plants begin to bloom together it can be a very amazing sight!

Propagation

The most common way to grow air plants is through propagation, or the creation of offsets, commonly known as "pups." While it depends on the conditions that the plants are kept in, most air plants will create offsets once they have finished the bloom process.

Bulbosa air plants produce a unique, multi-colored bloom from its center.

The *Tillandsia ixioides* showcases a lovely yellow flower when in bloom.

A close-up of the flower produced by the *Tillandsia neglecta* shows a myriad of purple and pink tones.

A *Tillansia streptophylla* spouts a red pup from under one of its leaves.

Tillandsia neglectas form an impressive clump.

Ionantha fuego air plants clump easily and will form spherical, colorful *Tillandsia* balls.

Tillandsia offsets grow in different ways depending on the species, with some creating pups around the base or root system, and others sprouting them from underneath one of their inner leaves (which serve as protection for the young *Tillandsia*).

Often a leaf will appear to be dead, but underneath that leaf a pup will be sprouting. (This is one reason why it's a good idea to proceed with caution when trimming or removing leaves). Some varieties, like the *funckiana*, will actually create pups on different parts of the plant, which creates a very unique look over time as the clump continues to grow. On average, air plants will create 1 to 3 pups after the blooming process. You can gently remove offsets from the mother plant when they grow to be about 1/3 the size of the mother. As mentioned earlier, if left un-separated from the mother plant, the offsets will continue to form a "clump."

Separated from their mother plants, these offsets, or "pups," will grow to maturity and be sold as individual plants.

GROWING FROM SEED

Growing *Tillandsia* from seed is a very long and arduous process. Once a *Tillandsia* blooms and is then pollinated, it emits a seed pouch that eventually will open up to release the seeds. Once released, the seed will need to be moved to a substrate, such as screening, to continue the growth process. It is important for the substrate to be able to retain some moisture, but not so much that the seedling sits in excess water, as this can promote fungus issues. The seeds should be kept moist and put in an area that gets moderate, indirect light and good air circulation.

The *Tillandsia brachycaulos x concolor* hybrid sprouts an offset from its base, underneath a leaf that has died.

It will take about a month for the seedling to begin to germinate, and the first few years of the growing process is very slow. Once the seedlings have reached an inch or so in height, the growing process accelerates a bit and the plants can be moved to an area where they get higher levels of light and receive care that is similar to a pup of the same size. This process can take some time, but the plants produced from seed tend to be bigger, healthier, and display better traits from generation to generation.

Inside the Operation:
A Look at a *Tillandsia* Nursery

In the fields of temperate Clermont, Florida is nestled a series of impressive greenhouses and shadehouse structures. It is the home of Russell's Bromeliads, a family-owned and operated *Tillandsia* nursery. Russell's has a long history as one of the country's major suppliers of *Tillandsia*, and in 2005, Mike Salvi, along with Debbie, his wife of 30 years, purchased the business.

Mike's longstanding passion for agriculture was what lead him to the acquisition of Russell's Bromeliads. After graduating in 1982 from University of Arizona with a degree in Agronomy, Mike worked in the hydroponics industry, and in 1984 opened an interior foliage business. His passion for agricultural innovation lead him to air plants.

"I suppose it was my love for growing things without soil that first attracted me to *Tillandsia*," he muses. "Once I discovered the unique characteristics of this plant, both aesthetically and botanically, I was hooked."

With millions of air plants under roof, Mike and his team at Russell's are well seasoned in the challenges that plant care can pose at a larger scale, even with a "low maintenance" plant like *Tillandsia*. One of the biggest considerations in terms of their care, he has found, is making sure they are provided adequate sunlight. "*Tillandsia* do best with bright but not direct light. This means that if you are growing them inside they should be receiving natural light through a window. Fluorescent light is also helpful."

Propagation is a big part of Russell's business, with thousands upon thousands of plants held solely for this purpose. When it comes to *Tillandsia* propagation, the key, Mike says, is patience. *Tillandsia* are an extremely slow to grow crop, and because of this, he says, the biggest challenge they face is keeping up with demand. "In order to supply our customers with exceptional quality, we are continually increasing production."

Because these plants are so slow growing, the unique challenge that they pose to those in the *Tillandsia* production industry is also a challenge that has helped to develop a deep-rooted sense of camaraderie and mutual appreciation among growers. This appreciation becomes apparent when talking about hybrids. Hybrid air plants are very popular, with a number of hybrid species in regular production. Hybrid air plants often showcase the strengths of the crossed species, and can even help ensure a more robust *Tillandsia* population. For example, the *capitata x xerographica* is a stately, exceptional hybrid that combines the heartiness of the *xerographic* with the propensity for offset production of the *capitata*. *Xerographica* air plants on their own are normally some of the slowest to produce pups, or offsets, but this beautiful hybrid readily produces them with proper care due to the *capitata* cross.

Although they grow a number of hybrid *Tillandsia* at Russell's, Mike personally hasn't developed any. But he holds much reverence for those who have. "Many of the hybrids which have

been developed were created by individuals who came before me. This process takes many years of dedication and to all who have been involved in this process we owe a great deal of gratitude."

While the slow growing nature of *Tillandsia* can pose a unique production challenge for nurseries like Russell's, their hardy and low-maintenance nature still make them an appealing plant for growers. Because they don't require soil, nurseries can keep a vast amount and store them both horizontally on racks, and vertically on walls, hooks, or hanging pots, maximizing the nursery space.

They are also, he says, fairly easy to keep pest-free compared to most types of plants, when given proper care (the right amount of light and water). A tip for plant owners, though: if pests such as spider mites or scale do develop, you can spray a diluted solution of ivory liquid soap (about 1 tablespoon per gallon of water) on the foliage to eliminate them and keep them under control. See pages 37–38 for more on what to do about pests.

While air plants were trendy in the 70s, they've re-emerged in recent years as a popular house plant among a younger generation who is discovering both the unique beauty and versatility of these plants. And with this new generation of air plant lovers, Mike has observed a renewed creativity when it comes to *Tillandsia* use and design. The fact that they don't require soil means that the display possibilities are truly endless.

Along with this surge in interest and creativity, these plants have also enjoyed a bit of an elevated status. "*Tillandsia* are used in a much more sophisticated manner than in the past," Mike observes.

And he is right. *Tillandsia* use is not just restricted to the typical house plant—they have made their way into the hands of many florists and event designers, not just as accent plants but in

many cases taking center stage in bridal bouquets, wreaths, centerpieces, and tablescapes.

Artists and artisans feature *Tillandsia* as muse, from living walls, to framed plant art, to hand-made goods that are created specifically with air plant display in mind.

• • • • • • • • • • • • • •

Tillandsia: the Versatile Houseplant

Because air plants don't require soil, they lend themselves to endless creative display and décor options. The vast variety of different *Tillandsia* species' shape, size, and overall look allows these plants to transcend many different decorating styles and functions. While air plants can live happily in their natural outdoor environments, they are prized houseplants as well, since they are relatively low maintenance, and their soil-less existence allows for them to be incorporated in so many different design and display options.

Choosing Your *Tillandsia*

There are many reasons to prefer one air plant over another, and aesthetic preference is certainly a driving factor in most people's decisions. However, as with all plants, when deciding which species of *Tillandsia* to select, your environment should be a major consideration. I've spoken to people in Arizona, for example, who can't seem to keep the mesic varieties happy, even with increased waterings. If you live in a super dry climate, a xeric *Tillandsia* will be a better bet.

Beyond your general climate, it's important to consider where you wish to display your *Tillandsia* when selecting the type of plant. While all species will need at least some degree of indirect light, certain silver-leafed types like the *xerographica* will want more light, and the fuzzy *tectorum Ecuador* will want very bright, even more direct light. Keep this in mind when selecting your *Tillandsia* species—if your room doesn't get a large amount of bright light, you should go for the mesic or semi-mesic varieties.

Limitations and Considerations

While there are so many ways in which you can incorporate air plants into your décor or creative projects, there are some key considerations to keep in mind with these plants. After all, low-maintenance or not, they are still living plants which require the proper care and environment to thrive. Here are some things to keep in in mind:

1. **Light:** this is one of the most important, and frequently overlooked, factors when it comes to choosing where and how to display your air plants. This should be a major consideration, as most varieties will need to be in an area that gets plenty of indirect sunlight. The xeric varieties (like *xerographica* and *tectorum Ecuador*) prefer more light and can take direct sunlight, so this should be taken into account when determining where and how to decorate with these plants. That bookshelf in the dark corner of your library might seem like the perfect place for your air plant, but it won't last too long in low light.

2. **Air Circulation:** as their name suggests, air plants derive their nutrients from the air, so it's important to make sure that wherever you display these plants, they have good air circulation. Any terrariums or containers should be open to allow for air circulation. Additionally, air plants tend not to do as well when near a heat or AC source, as this can cause the plant to dry out.

3. **Attachment Method:** While there are many ways that you can showcase *Tillandsia*, you don't want to use any method of attachment to an object or display that could damage the plant. For example, wire may be wrapped around the plant base, but not pierced through the base or leaves.

You can also use a plant-safe glue, although it's important to keep in mind that this can put restrictions on how the plant can be cared for (it won't be as easy to soak a plant that is permanently attached to a wreath with glue).

4. **Moisture:** You never want your air plants sitting in or holding moisture, as this can cause the plant to rot. This means:

- Instead of living, damp moss, you might use preserved moss.
- Gravel, rocks, or sand should be fully dry, and these plants (with the exception of a couple varieties) should never be planted in soil.

- Air plants should be allowed to dry fully with no water standing in leaves before placing the plant back into a container or display.
- If misting the plants, make sure you're not misting them into a container that could hold the moisture against the plant and not allow it to easily evaporate, causing rot or fungus, which are, unfortunately, irreversible in most plants.

5. **Toxicity:** There are a few materials that are toxic to *Tillandsia*. Obviously, avoid contact between these materials and the plants or they could severely compromise plant health. These are:

- Copper: while copper can be beautiful to decorate with, it is toxic to *Tillandsia*. This means you shouldn't use copper holders, wire, or even pressure treated lumber (the chemicals used to treat the lumber contains copper).
- Boron and Zinc: These are both also toxic to *Tillandsia*, and it's important to avoid any fertilizers that contain these elements.

Displaying Air Plants on the Wall

Art can truly transform a space; it can personalize, perk up, and complete. Because air plants don't require soil, they lend themselves perfectly to vertical displays and artistic wall hangings. Their unique shapes, curling leaves, and other-worldly appearance let them serve as living art, and their many different looks provide endless possibilities for complimenting one's personal style.

We have framed art and photos in every room of our home—we like to bring back art as souvenirs from our travels so that our walls become a sort of evolving album for our memories. But the piece that we get the most comments on by far is the frame that holds many of our air plants.

This thrift store frame has the glass removed and tacks stuck haphazardly around the perimeter. The string is looped around the tacks and pulled taught in a zigzag formation. Small air plants can be carefully tucked between the strings and easily removed when it's time for watering.

Air Plant Frame

The rustic beauty of natural wood frames a grouping of air plants on a well-lit living room wall. This frame was actually dug out of Ryan's father's attic, making this piece extra special for us to display in our home.

We constructed this frame by using heavy gauge wire crossed in a grid and wound tightly around screws mounted into the frame. You could easily construct a similar frame with a staple gun and chicken wire. The wire will provide the structure on which you can gently curl, nestle, or tuck the *Tillandsia*. The size and spacing of the grid you create (if using wire) or chicken wire will determine the look of your frame and also which types of *Tillandsia* will best be displayed in it. This creates a beautiful living piece of art—modern and unexpected.

Take the *Tillandsia* down gently to water them, and allow them to dry fully before returning them to the frame. Make sure the frame is on a wall that gets plenty of indirect sunlight, as this will help ensure your framed living art continues to thrive.

We took an old wood frame that had sat for years in a family attic and gave it new life with air plants.

Adding Air Plants to a Floral Arrangement

It probably should come as no surprise that our home is filled with air plants. But that doesn't mean that we are an "air plant exclusive" household. In fact we enjoy filling our home with all

We constructed a simple, minimalist and modern bouquet with several sprigs of florals, a sprig of eucalyptus, and a large *Tillandsia* perched on top of the vase of our bar. The vase was a wedding gift from my mother, a hand-painted piece from Okinawa, Japan.

A blooming *Tillandsia* is perched with a simple arrangement of bright pink tulips for a beautiful spring table accent.

sorts of greenery and florals. There is something so inviting about a thoughtful arrangement of fresh flowers, and I absolutely love the character that air plants add to an otherwise beautiful but perhaps a bit blasé floral bouquet. Likewise, while *Tillandsia* are beautiful on their own, they can also be complimented by the colorful elegance of flowers.

It's very easy to incorporate air plants into a floral arrangement. Simply create your arrangement, and then select air plants that can sit amidst the flowers and greenery. The important thing is to make sure that they are propped up on the dry part of the floral arrangement and don't sit in any water.

The round, wide stance and stiff leaves of the *Tillandsia* allow it to perch atop a bouquet of spring blooms.

Then, once the flowers have expired, simply remove the air plants and find a new home for them! That is one clear advantage of air plants over cut flowers: they live on way past the occasion for the arrangement.

Tillandsia: The Entertaining Plant

Perhaps the most wonderful thing about *Tillandsia* when it comes to entertaining is their keen ability to transition from casual, every-day accents to show-stopping, special occasion center-pieces. Used on their own in simple groupings, incorporated with florals, or arranged in special table-top terrariums, *Tillandsia* make the perfect and easy compliment to your table, whether for a casual weeknight dinner with your family, or a special spread to celebrate and impress.

A simple arrangement of air plants adorn a sunny dining table. The unexpected pairing of elegant crystal candlesticks with organic and free-flowing *Tillandsia* make for an un-stuffy spread that is pretty without trying too hard.

Use a mix of small and large *Tillandsia* nestled on driftwood as a rustic-modern centerpiece perfect for every-day dining and last minute entertaining.

The incorporation of air plants on a wood farmhouse table with seaglass-blue vases balances natural elements with elegance and ease.

Create a beautiful, summery centerpiece by pairing a beachy terrarium with a few additional *Tillandsia* accents.

Throwing a party? Use small air plants as place settings, which double as sweet favors for your guests to take home! It's as thoughtful as it is beautiful, and you get to spread your love and passion for *Tillandsia* with your friends and family!

Tillandsia in Outdoor Living

We relish in spending time outside. Living in Florida, we're fortunate that we have temperate weather year round so we're never shut in (although we're perhaps a little too warm for a couple months). This allows us to display *Tillandsia* in our outdoor spaces year round as well as indoors. Because *Tillandsia* in general prefer temperatures above 50 degrees F, outdoor air plant displays

won't be possible if you live in a cooler climate, except for perhaps in the late spring and summer months.

An elegant space that blurs the line between indoor and outdoor living, this screened patio features a rustic-modern centerpiece of globes inspired by natural elements, and a thoughtful scattering of *Tillandsia*.

Holiday Decorating with Air Plants

A modern holiday tablescape of metallics, sparkling candles, and air plants in varying shades of green set a festive but not too kitschy mood for a cozy Christmas dinner.

An autumnal centerpiece: Mix an array of air plants with a selection of gourds and pumpkins to create a spread that is a perfect compliment to cooling temperatures, falling leaves, and fall festivities.

I absolutely love using *Tillandsia* in holiday decor, primarily because holiday decorations can be a bit kitschy, and the incorporation of air plants seems to both ground everything a bit as well as provide a fresh update—a bit of originality. And with many shapes, sizes, and shades, *Tillandsia* truly bring endless customization options to your holiday decorating.

From festive table settings, to ornaments, to wreaths, to mantle decor and beyond, *Tillandsia* can bring some organic yet totally unexpected personality to your holiday spread.

The Holiday Terrarium

The beauty of this easy, DIY project is that you can easily change out the terrarium fillers to shift with the seasons and compliment your favorite holiday. This terrarium works particularly

A bright mix of *Tillandsia* and reindeer moss in a simple glass globe, hung by red satin ribbon, makes for a festive terrarium that can be displayed year round. Add some red or pink glass stones or sand for Valentines Day and give a gift that lasts far longer than a bouquet of roses.

well for Christmas, but the red ribbon could easily be switched for blue or silver, and substitute a blue colored moss or stones in place of the green moss for a Hanukkah terrarium.

You will need:

1 glass vessel, globe, or terrarium

Several small air plants (the red color of the *brachycaulos abdita* can be a great addition to a Christmas-themed terrarium)

White sand as a base

Green preserved reindeer moss

To assemble: simply start by filling the terrarium or globe base with white sand, perhaps to ¼" depth. Add pieces of preserved green reindeer moss (you can use any type of preserved moss here, but not living moss as this can retain moisture against the plants and cause rot). Then, one by one, gently slide in and position your air plants, nestling the base of the plants slightly into the moss and sand.

Tillandsia can also be a great compliment to other seasonal decor. I love finding other natural elements of each season to create arrangements that are appropriate for the time of year, but not overtly themed.

Displaying Air Plants: Shelf Life

Tillandsia can make a gorgeous and unexpected addition to a bookshelf, floating shelf, or any type of shelf that gets sufficient light. Because they don't require soil, air plants can be displayed in nearly any container or on their own, simply seated shelf-top.

Place the striking *Tillandsia fuchsii* atop a selection of favorite books for a funky accent.

Megan George (of The ZEN Succulent) showcases a beautifully-styled terrarium, featuring air plants, natural driftwood, moss, and sand, alongside an eclectic mix of art, vintage and sentimental items, and books. >>

Displaying Air Plants: The Easy Terrarium

Tillandsia terrariums are incredibly easy to make as there is no "planting" involved. The possibilities are endless when it comes to styling your terrarium, but the biggest consideration when it comes to a hanging air plant terrarium is that all terrarium materials stay dry, and the plants don't ever sit in an environment where moisture is held against them. It is also important that you remove the plants from their terrarium environment to water them, and allow them to fully dry before putting them back in their hanging home.

Megan George from The ZEN Succulent constructed this terrarium with natural wood and shell accents that create an organic environment for the *Tillandsia*.

Terrariums can be constructed and displayed in hanging vessels, or seated tabletop. Again, the most important factor to where you display your terrarium is light. You also want to make sure that whatever terrarium you select is open on one side, or at least partially open, to allow air to properly circulate.

CHOOSE YOUR TERRARIUM

Select the vessel you want to build your terrarium in. This could be a simple glass globe, a teardrop shape, or a geometric glass terrarium that is edged in modern metal. Do you want it to hang or do you want it to sit atop a tray on your coffee table? Do you want one large statement terrarium, or a series of small hanging globes strung at different levels?

You don't want to put *Tillandsia* in a container that is completely closed for a prolonged period of time, as this will suffocate and eventually kill the plants. Be sure your terrarium vessel allows for plenty of air flow.

Also keep in mind that the size of the terrarium opening will determine to a large extent how easy it will be to set up your terrarium, and even which plants you should use. Generally, the larger the opening, the easier it will be to situate the plants and other materials. For smaller openings, you may want to enlist the help of small tongs or tweezers.

The size of the terrarium opening will also affect the care that is needed for the plants. A more closed terrarium will naturally trap more humidity, so you may find you need to water the plant less in a terrarium with a smaller opening.

An alternative to a closed terrarium: take a simple dish and some glass stones and nestle the plant on top. The clean contrast of white ceramic and black glass stone creates the perfect setting for a blooming *Aeranthos*.

Note: It can't be said enough that one of the most common (and sadly irreversible) detriments to *Tillandsia* is prolonged exposure to excess moisture, which can cause mold, fungus, or rot. This is why it is generally not recommended to water or mist your plants while they are in their terrarium, as water can collect in the terrarium and sit against the plants. I also highly recommend that after watering plants, they are allowed to fully dry before you put them back in their home where airflow may be a bit more restricted.

Your terrarium container doesn't necessarily need to be glass (in fact, in a later section we'll highlight a unique fiber artist who makes terrariums out of felt), but many will prefer glass because it will afford a full 360 degree view of the *Tillandsia* environment. Glass terrariums will also generally allow for more light to be let in, and as we've discussed, light is critical. If you select an opaque container, make sure the opening is big enough to let in plenty of light, or that the plants can be positioned in such a way that they are receiving light.

A gold pyramid, open on one side for good air flow, is a glamorous home for a bright red *abdita* on a simple bed of white decorative sand.

Also, while copper-framed geometric terrariums can be beautiful, remember that copper is toxic to air plants. You may want to double check that if your terrarium container has a metallic component to it, that it is not made from or with copper.

Some good terrarium container options:

- Hanging glass globes
- Metal-framed geometric terrariums
- Mason jars
- Wide glass bowls or vases
- Large wine glasses
- Glass pitchers

Sometimes simple is stunning. This glass globe holds a single blooming *Tillandsia aeranthos*.

CHOOSE YOUR PLANTS AND THEIR ACCOMPANIMENTS

Keep your terrarium simple and striking by featuring a single, larger air plant in a simple glass globe. This can be a beautiful look alone or done in a series of three for an easy, minimalist, organic installation.

Add sand or crushed stone as a base, with a bit of preserved moss.

The striking and rare *Tillandsia victoriana* shines in a simple mason jar terrarium.

Beach-Inspired Air Plant Terrarium

You'll need:

2 to 3 small air plants. Recommended: *ionantha* and *bulbosa* varieties.

Glass globe or bowl (hanging or tabletop)

White sand or crushed stone (the amount will depend on vessel size, enough to create a base)

Green or natural preserved moss

Small to medium-sized seashells

Assembly:

Note: If you plan to hang your terrarium, it's best to assemble your terrarium while it's situated on a tabletop, and then carefully hang it once assembled. It's also helpful to use tweezers and even chopsticks to help arrange the different components once they are in the terrarium, especially if there is a small opening.

1. Add sand or stone to the bottom of the vessel.

2. Add moss and shells in a natural-looking pattern. The moss can be used in larger groupings to mimic dunes or broken up to be reminiscent of seaweed.

3. Situate seashells and air plants as desired.

And just like that, a bit of the beach has been added to your space (even if you live in the Midwest and it's February).

A Terrarium Artist: Megan George and The ZEN Succulent

Megan George, terrarium artist and owner of The ZEN Succulent, a modern terrarium & crafts shop based out of Raleigh, NC, has truly established herself as a terrarium artist. She is also author of *Modern Terrarium Studio*, which features terrarium projects using succulents, cacti, and air plants.

Characterized by an eclectic blend of natural and locally-sourced elements in creative vessels, Megan's terrarium creations and arrangements highlight the natural beauty and modernity of *Tillandsia* shapes.

As a business school graduate, Megan has successfully combined her passion for lush greenery with her aptitude for social media marketing and her education to grow The ZEN Succulent into a nationally and internationally known business that draws customers all over the US, the UK, Australia, Switzerland, and beyond. Her terrariums have appeared online on *HGTV Gardens, Mental Floss,* and *Etsy Editor's Picks* and in The Fayetteville Observer *Sunday Life*.

Megan displays a series of hanging terrariums in a sunny window, accented by *Tillandsia usneoides,* commonly knows as Spanish moss.

Megan draws inspiration for her terrarium creations from her natural surroundings in her home in Raleigh. Her love of urban gardening and delight in color comes through in many of her creations, and she is active in her local community, teaching and leading workshops that share her modern take on traditional plant terrariums.

• • • • • • • • • • • • • •

An Alternative to Glass: Fiber Artist Justine Moody and the Unexpected Terrarium

Perhaps one of the clearest illustrations of the versatile nature of the *Tillandsia* is the unique and artistic crafts that they inspire. Fiber artist Justine Moody shows this in her hand-felted planters, created specifically with these soil-less plants in mind.

Based out of Long Island, NY, Justine creates one-of-a-kind felted planters from natural fibers. While some of her planters feature small pops of color, they predominantly use a soft white and grey color palette that is reminiscent of birch bark, further tying her work to the natural surroundings that inspire it. Because air plants don't require soil, these planters are uniquely suited to showcase them.

While her work is nothing short of innovative, and certainly would be considered modern in aesthetic, Justine revels in the fact that her medium allows her to connect on an artistic and fundamental level with the tradition that has inspired it. "Working

<< Some of Justine's felted terrariums are open on top, like this one, which allows the tentacle-like leaves of the *caput-medusae* to twist out the top.

with natural fiber allows me to embrace with and connect to centuries of tradition in a modern art form," she muses.

Her work with natural fibers also allows her to connect to the organic world in a way that many craft forms could not, and she is conscientious of the sustainable nature of her medium. "I'm mostly drawn to the organic renewability of natural fibers and have a vast respect for the history of the craft. I love the tactility of the wet felting process and it's eternal connection to nature and the animal it once belonged to, a simple yet magical transformation."

The incorporation of air plants into her art, she says, only draws her work closer to its natural source.

Justine uses a variety of small- to medium-sized plants in her felted planters, for an overall look that is both whimsical and grounded, modern and organic.

While she adores so many varieties of *Tillandsia*, Justine says she is partial to the mysterious look of the *pseudobaileyi*, and also loves to showcase the *aeranthos*, especially when it is in bloom.

Justine's work in itself draws many similarities to the *Tillandsia* it showcases: full of history, yet freshly modern—organic, yet otherworldly.

Displaying Air Plants: Natural Elements

While *Tillandsia* all have unique shapes of their own, many of them can be reminiscent of other plants and creatures. It is perhaps because of this, and of course due to their versatility and low maintenance nature, that they can look great displayed with other natural elements.

DRIFTWOOD

The twists and turns of natural driftwood create perfect nooks in which to situate air plants. These blooming *Tillandsia stricta* plants sit atop a piece of driftwood. The nooks in the driftwood situate the plants perfectly, with no need for glue. This would be a gorgeous piece for a windowsill, dresser top, or dining table centerpiece.

SEASHELLS

A popular yet delightful way to showcase air plants is in natural seashells.

Create a funky sea creature-esque character by pairing a *Tillandsia caput-medusae* with a large sea urchin shell.

Place small *Tillandsia ionantha* plants in small pin sea urchin shells for a fun and colorful accent to your decor.

Place blooming air plants in the nooks of driftwood to achieve a look of them growing out of the log.

Several *Tillandsia bracycaulos x concolor* perch atop a piece of driftwood.

A bright, modern bird sprouts an *ionantha* "feather."

This gilded porcelain hedgehog was a gift from our cousin, and it seemed the perfect showcase for a blooming *Tillansia stricta*.

The Whimsical Air Plant

There is something just plain fun about taking a cute animal-shaped vase or figurine and giving it some actual life with an air plant. And with an increasing number of younger air plant enthusiasts, we see all sorts of examples of air plants being used in fun and non-traditional planters.

I love taking a funny little animal-shaped dish and adding an air plant—it's a great way to animate the inanimate and inspire happiness.

Tillandsia: Wearable, Natural Art

Air plants are emerging not just as a trendy houseplant, but as a plant that can be used as a fashionable accessory. From air plant jewelry, to wearable wreaths, *Tillandsia* are being used more and more in wearable art.

Some air plants, like the *funckiana*, can take on a wearable form without any additional help. This blooming and mature *funckiana* was so long and curved that it created a great bracelet (although it is a bit reminiscent of a caterpillar).

And it's not just humans that can sport these fun plants. *Tillandsia* can make a beautiful accent to a wreath for a "flower dog" (or for any time you want to dress up your pooch).

Molly, a 5-year-old American Bulldog Catahoula mix, sports a *Tillandsia* wreath that we constructed for a client's wedding. The wreath was ultimately worn by their dog who walked them down the aisle for their wedding. For this wreath, we used a base of a biodegradable Styrofoam ring so as not to have something too heavy for the dog, and wrapped it in shades of purple ribbon to compliment the couple's wedding colors. We then attached various small *Tillandsia* using a plant-safe glue at different angles.

Tillandsia in Retail Displays

As *Tillandsia* have grown in popularity, and emerged as the "it" plant, it's becoming increasingly easy to find them at local brick and mortar retailers. Many home decor and lifestyle boutiques sell *Tillandsia*, even ones that would never consider themselves a "plant store," because of how relatively easy they are to keep. They are also used to style retail displays, adding an organic element that doesn't need to be planted in soil and so can be situated in many different ways to compliment the product display and set the mood.

In picturesque Charleston, SC, we visited Candlefish, a brilliantly curated shop featuring candles, gifts, and home accents from independent makers. They not only sell air plants, but use them to showcase their other products, like hanging handmade hemp cord baskets.

They also use them around their shop to accent product displays.

• • • • • • • • • • • • • •

Air Plants in Love: the Modern Wedding Plant

With the evolution of the *Tillandsia* in modern design in recent years, these plants have started to transcend their traditional role as houseplants to infiltrate new disciplines. Perhaps somewhat ironically, air plants have emerged as a fresh staple in one of the most traditional institutions: the wedding.

Air plants have surged in popularity with professional floral designers, event designers, and DIY brides for a variety of reasons, and although it's tempting to dismiss them as trendy, the uniquely versatile nature of these plants ensure that they will not be a fleeting fad.

The Air Plant Wedding Bouquet

With their twists and turns, spikes and curves, air plants provide the unexpected element to traditional florals, or can be used by themselves or with succulents to create a more rustic-modern bouquet. It all depends on the style of the event and the couple (and their budget), but there is no shortage of options when it comes to using *Tillandsia* in a bridal bouquet.

A PERFECT BALANCE

The mix of more traditional florals with air plants can create a beautiful juxtaposition that is both modern and timeless. In this bouquet, the lush white giant hydrangea and delicate baby's breath are given an unexpected update with the stately *Tillandsia xerographica*, and bright accents of crespedia.

The *xerographica* in this bouquet brings a bit of wild organic madness to an otherwise fussy flower. The harder edge of the tilly's leaves ground the romantic florals to compliment a wedding style that is elegant, but doesn't take itself too seriously.

Tips and Considerations

While *Tillandsia* are a striking addition to a floral bouquet, the larger ones like the *xerographica*, which is also one of the most often used varieties in bouquets, can be a challenge to anchor to the bouquet. They don't have a "stem" and their wide bases can be weighty depending on the size of the plant.

A single *xerographica* serves as the perfect bouquet for a less-than-traditional bride.

It's a good idea, in addition to using wire, to anchor the base of the plant to a floral stake with a bit of plant-safe glue. This will allow the plant to have a "stem" that can then be gathered with the stem of the other flowers in the bouquet.

Note that it's best to avoid piercing through the base of the plant with wire if you plan to keep the plant after the wedding. Using a soft covered aluminum wire works well as you can easily bend it around the base to help secure the plant.

MODERN SIMPLICITY

Air plants can certainly enhance the traditional floral bouquet. But if your bridal style is simpler (and likely less expensive) or more modern, air plants can stand alone.

Due to its impressive size and stately, global shape, the great Queen of the Air Plants, the *xerographica*, can be carried on its own as the entire bouquet. This is appealing to a less-than-traditional bride who wants to stand out from the norm.

It's also appealing to the practical bride who doesn't want to (or can't) spend hundreds of dollars on a bridal bouquet, only to throw it away, or at best dry and preserve it. The simple *xerographica* can be kept for years and years as a cherished houseplant to remind the couple of their wedding day long after the vows have been said.

Not sure about a *xerographica* for your bridal bouquet? How about using them as simple alternatives to your bridesmaids' bouquets! After the ceremony, these can be used in the wedding centerpieces or to accent other florals or decor, and they can be given to your wedding party or guests at the end of the night as a special gift.

Wearable *Tillandsia*: Boutonnieres & Corsages

Smaller air plants can be used for boutonnieres, either on their own or with some simple floral accents. *Tillandsia* varieties that have more narrow bases tend to work best for boutonnieres, as the narrow base will be easy to wrap with floral tape against a boutonniere stake or with other florals.

The type of *Tillandsia* you select for the boutonniere or corsage will also, of course, determine the overall style of the piece, and as *Tillandsia* come in so many different shapes and forms, there is no shortage of options for creating a unique piece that perfectly compliments your wedding style.

The Air Plant Boutonniere

If you're looking to be more hands-on with your own wedding details, the boutonnieres can be a great project to take on, even without a background in floral design. While there are a number of methods to assembling a boutonniere or corsage, a combination of floral wire and tape seem

to work best when it comes to actually securing the air plant to the rest of the floral materials for a "mixed media" boutonniere.

If you are planning to include other elements in the *Tillandsia* boutonniere, such as a sprig of greenery or baby's breath, one of the best things you can buy are the miniature floral stakes which have a small strand of floral wire already attached to an end. Specifically for boutonnieres and similar projects, these can be found at most crafts stores or shops that sell floral or hobby supplies. You can then use this stake and wire (some additional floral wire might be a good idea) to wrap the narrow base of the plant, along with the stem of the flower or other elements you are incorporating. Then, once assembled in floral tape, wrap the entire base to

Here the cushiony shape of the *xerographica* makes a perfect pillow for a shot of the rings.

The stately *xerographica* adorns a window in The Sanchez House bathroom, providing quiet contrast to the worn historic architecture.

finish. You can leave the base with simply the tape to be pinned, or wrap some ribbon or twine around the base that compliments your wedding color palette and style.

If you are going for a simple, streamlined look (and a lot less work), the personality of your air plant is strong enough to allow it to stand on its own as the entire boutonniere, no additional flowers or greenery required! For a super-simple air-plant-only boutonniere, you can simply wrap the plant base with floral tape (and a ribbon or other material if you so choose), enough so that the pin will stick through the tape and/or ribbon and not through the plant itself.

Air Plants: The Living Favor

Because most *Tillandsia* aren't rooted in soil, they can be presented as gifts in unique ways that are easy to transport to any wedding or event venue. Fun on their own, or presented in a seashell, gift box, terrarium, or other vessel, they make for a great party favor. An added benefit: they can serve double duty as both favor and decor when used at the place setting of your guests.

Wedding *Tillandsia* Decor

Just as air plants are versatile in home decor, so too are there endless ways to adorn your wedding with them. As they have surged in popularity in recent years, professional wedding florists and stylists have been finding innovative and amazing ways to work these fun and funky plants into their design.

For the earthier couple, air plants can completely replace florals for an organic and grounded style. Air plants can be arranged on their own as centerpieces, or featured in terrariums that are constructed to suit the style of the event and space itself.

Tillandsia can also be placed in terrarium globes and hung, either in the event space, on the wedding arch, or both! In fact, some of the prettiest and most romantic wedding arches making their way across Pinterest boards today feature *Tillandsia*.

Tillandsia can also be added to simple floral arrangements to add a bit of unexpected delight and a modern edge.

Many photographers will use air plants as styling props, and when air plants are incorporated into the wedding day, they tend to make it into the wedding photographer's shots.

A Fresh Accent to a Historic Wedding Venue

Our wedding was the definition of DIY. Admittedly averse to the "wedding theme" concept, we wanted a laid-back and authentic yet elegant style for our celebration. We chose The Sanchez House in St. Augustine, FL for our setting, and the historic charm of this 1800s Spanish-influenced structure allowed us to do very little in terms of overall decorating. In fact, keeping the decor simple was a necessity with this place, as it would have been a shame to overshadow the beauty and history with overly-ornate arrangements or a kitschy theme.

We chose to use *Tillandsia* to accent and enhance the existing beauty of the house. While it might seem counterintuitive to use such a modern-looking plant in a historic setting, it worked perfectly for our style as a couple, as we like to blend both vintage and modern elements in our home. When we started planning our wedding, our vision for the whole event was that it feel less like an event and more like a cozy and comfortable celebration.

The air plants on our wedding day also made for a fun project for friends and family who helped us put it all together. One of my bridesmaids had the idea to utilize a multi-tiered candle holder that sat in the fireplace as a showcase for air plants. I loved the idea, as it brought a bit of greenery to an unexpected area. We also sprinkled air plants on the mantle above the fireplace.

Tillandsia, perhaps not surprisingly, were featured not just in our decor but in the bridal bouquet and boutonnieres, and their common yet understated incorporation throughout the day helped to tie all the details together in a cohesive, though not overtly orchestrated, package.

• • • • • • • • • • • • • •

Finding Your Own *Tillandsia* Love

With the endless possibilities that air plants present, and their low-key nature when it comes to care, it's not surprising that they continue to grow in popularity with so many different types of people.

Apartment dwellers can appreciate their smaller scale and soil-less existence, easily adorning even the smallest of urban dwellings with a bit of greenery and life.

Artists and artisans find inspiration in their unique, funky, sometimes kind of crazy shapes and silhouettes.

Professional floral designers and amateur DIYers alike revel in their versatility, and their ability to bring a bit of unexpected flair to even the most traditional setting.

We love *Tillandsia* for so many reasons and ways, and seem to find a new way to love them every day.

How about you?

Credits and Resources

Text by Ryan and Meriel Lesseig

Photographs:

All photography by Ryan and Meriel Lesseig with the exception of the following contributors:

Abigail Gehring: page 98

Benj Haisch: pages 136, 143

Justine Moody: pages 122–124

Leslie Hollingsworth Photography: pages 132, 135, 138, 141–143, 144–149

Megan George, The ZEN Succulent: pages 112, 115, 121:

Research Assistance:

Tara Moran

Resources:

Tillandsia: The World's Most Unusual Air Plants by Paul T. Isely, Botanical Press, 1987

Air Plants: The Curious World of Tillandsia by Zenaida Sengo, Timber Press, 2014

Online:

Rain Forest Flora: www.rainforestflora.com

Bromeliad Society International: www.bsi.org/new

Jack's Florida Bromeliads: www.jacksbromeliads.com

Florida Council of Bromeliad Societies: http://fcbs.org

Bromeliads.info: www.bromeliads.info

Tropiflora: http://tropiflora.com

Bromeliads in Australia: www.bromeliad.org.au

Index